A Train with a Track in the Box

A Train with a Track in the Box

Charlotte Harris

Copyright © 2017 by Charlotte Harris.

ISBN: Softcover   978-1-5434-8516-5
          eBook   978-1-5434-8517-2

All rights reserved. No part of this book may be reproduced or transmitted in any form or by any means, electronic or mechanical, including photocopying, recording, or by any information storage and retrieval system, without permission in writing from the copyright owner.

Any people depicted in stock imagery provided by Thinkstock are models, and such images are being used for illustrative purposes only. Certain stock imagery © Thinkstock.

Print information available on the last page.

Rev. date: 04/21/2017

To order additional copies of this book, contact:
Xlibris
800-056-3182
www.Xlibrispublishing.co.uk
Orders@Xlibrispublishing.co.uk
688213

ONE

The most amazing thing to happen to me—quite recently in fact—was to

Charlotte Harris

be included in a group of men and women scientists at the LHC (Large Hadron Collider) at CERN (European Organization

A Train with a Track in the Box

for Nuclear Research), Geneva, and to be able to help prove the existence of the so-called God Particle which establishes mass

Charlotte Harris

and makes things
heavy due to
inertia or a pull
through a field
as if dragging
a particle or
anything else
though molasses.

A Train with a Track in the Box

That summer, I was studying under the watchful eye of a certain Dr Todd B. Huffman, a professor in his field and a tutor for many other

Charlotte Harris

subjects that you could wish for. We teamed up for talks in the kitchen, where most of my ideas were hatched, towards producing

A Train with a Track in the Box

a paper sometime later when they were required for a journal and kept up to standard just in case. It was on a quiet, rainy afternoon that

Charlotte Harris

Dr Huffman and I were watching a model train going round a track in a halting monotone that my dear friend said to me, 'On the

A Train with a Track in the Box

train, you're going fast. From the embankment the train runs slow.

If you try to measure relative time, it is only the

Charlotte Harris

speed of light c that we know.

The constant **c** is the speed of light.

The train becomes heavy;

A Train with a Track in the Box

the track might

brake

Far from slowing

out of sight

The train speeds up. Which track do you take?'

I of course did not know how to respond and I had to admit that

A Train with a Track in the Box

I did not know. After a while the conversation turned to mathematics.

'Did you know that if a times b

equals c (and a and b can be any positive whole number) then a divided by b times c equals a squared?'

A Train with a Track in the Box

$a \times b = c$

$a/b \times c = a^2$

I had to say that I did know and that this was in fact part of my research.

Charlotte Harris

'Did you know that A raised to the power of b divided by c equals E and this can be reversed in E raised to the power of c divided by b?'

A Train with a Track in the Box

$A^{(b/c)} = E$

$E^{(c/b)} = A$

Dr Huffman explained that x is the power to which the base is raised to produce the exponent. The resultant raised to a power is also an exponent.

B^x is an exponent.

A Train with a Track in the Box

I nodded my agreement. 'I'll write down my calculations,' I said. 'I think that you will find these very useful.'

TWO

'The exponents we already know are forms of the grand exponents equation.

A Train with a Track in the Box

$$A^x + A^y =$$

$$A^{(x+y\,/2)}$$

$$((A^{(x-y\,/2)} +$$

$$A^{(y-x\,/2)})$$

'The subtraction is very similar.

$$A^x - A^y = A^{(x+y/2)}(A^{(x-y/2)} - A^{(y-x/2)})$$

'From this, we learn that

A Train with a Track in the Box

$A^x + A^y =$

$(A^{(y-x)} + 1) \times A^x$

$A^x - A^y =$

$(A^{(y-x)} - A) \times A^x$

'The most thoughtful equation is still

to come. The power n is the subject and can be any whole number suggested between x and y. The power n takes part in each term

A Train with a Track in the Box

on the right in the equation.'

$A^x + A^y = A^n$

$(A^{(x-n)} + A^{(y-n)})$

$A^x - A^y = A^n$

$(A^{(x-n)} - A^{(y-n)})$

Charlotte Harris

'These three calculations are in addition to other equations learned already.'

A Train with a Track in the Box

$4^3 + 4^7 = 4^{(3+7)/2} \times (4^{(7-3)/2} + 4^{(3-7)/2})$

$= 4^5 (4^2 + 4^{-2})$

$= 4^5 (16 + 1/16)$

$= 16(1/16) \, 4^5$

$= 16448$

Charlotte Harris

$7^8 - 7^2 = 7^{(8+2)/2} \times (7^{(8-2)/2} - 7^{(2-8)})$

$= 7^5 (7^3 - 7^{-3})$

$= 7^5 (7^3 - 1/7^3)$

$= 7^3(1/7^3) \, 7^5$

$= 5764752$

A Train with a Track in the Box

$5^{12} + 5^4 =$

$5^8 (5^{(12-8)} + 5^{(8-12)})$

$= 5^8 (5^4 + 1/5^4)$

$= 5^4(1/5^4) \, 5^8$

$= 244141250$

Charlotte Harris

$8^{10} - 8^7 =$

$8^9 (8^{(10-9)} - 8^{(7-9)})$

$= 8^9 \times (8 - 8^{-2})$

$= 8^9 (8 - 1/8^2)$

$= 8(-1/84) 8^9$

$= 1071644672$

THREE

A Diary Entry

A strange thing happened to me today as I

Charlotte Harris

was writing an equation. I was writing it up on the board— the board along the corridor— where there was no one to see

A Train with a Track in the Box

my shaky hand and the piece of paper folded into quarters, where my latest equation was written elaborately, almost in a scribble. If

Charlotte Harris

by some turn of fate someone did see me writing on the blackboard available for strong statements and hidden truths maintained

A Train with a Track in the Box

asi diologue, I would pretend to be wiping the board clean. I may in the event have to do so anyway. My puny statement was

Phil (r)=....[]

And

P (r) =...[]

The radical probability density of the electronic

A Train with a Track in the Box

being somewhere in the atom.

My thinking was this: the atom does not have a sharp boundary so an approximation would be as good as any other answer which may be incorrect by trying to pin it down to a single answer.

A Train with a Track in the Box

No sooner

had I scrawled

my equation

in my neatest

blackboard

handwriting

than an answer

appeared at

the top of the blackboard.

'The probability of finding the electron within volume dV is ψ squared dV.

A Train with a Track in the Box

Probability = ψ^2 dV.

'The radial probability density P (r) holds that P(r) dr is the probability of finding the electron within the spherical shell from r to r + dV. The volume of a shell of radius r and thickness dr is dV.

A Train with a Track in the Box

$$dV = 4\pi r^2$$

$$p(r)dr = \psi^2 (4\pi r^2 dr) = \psi \, dV$$

'The radical probability density is

$P(r) = 4\pi r^2 \phi$

Then for the first state

$P(r) = 4r^2/r_0^3 \, e^{-2r/r_0}.$'

I was excited and thrilled to receive

A Train with a Track in the Box

a reply from an invisible person. I whispered hello but there was no answer. I coughed and pretended that I was going to write more and then the writing stopped. Unlike my poor equation, this writing was very clear and the brevity of my work became lucid.

A Train with a Track in the Box

Five whole minutes went by with students walking along the corridor glancing at the board and sometimes stopping to read. Somehow I overcame my shyness on both accounts. I whispered 'Who is it?' in a loud stage whisper. I thought that this

A Train with a Track in the Box

was mysterious and very exciting. There was a folded piece of paper on the blackboard tray. I opened it and obediently wrote the script up on the blackboard.

'The magnitude of the spin, angular momentum S of the electron is determined by its spin quantum number S=1/2

S= $\sqrt{S(S+1)}$ and hbar= $\sqrt{3}/2$ hbar.'

Charlotte Harris

Then the voice from the invisible man said, 'Can you attempt a solution?' I said yes, this was very easy. In a magnetic field the

A Train with a Track in the Box

z component can only assume two values.

I wrote

$S_z = M_s \hbar$

and calculated the answer.

I came up with the same answer twice: 0.866025403 and 0.866025403. 'Can I go now?' I dared ask.

A Train with a Track in the Box

'Yes', he replied, 'just wipe the board for me.'

I was very keen to leave $Sz=Ms$ hbar on the board. It was very profound

Charlotte Harris

and I wiped the blackboard very clean leaving the electron behaviour untouched and went home. Next entry, tomorrow.

FOUR

'So what is the answer?' I asked and Dr Huffman smiled.

Charlotte Harris

'I'll tell you in a minute,' he said.

'If the train runs at the speed of light and becomes infinitely heavy, it does not matter

A Train with a Track in the Box

which track it is on because it would achieve a standstill relative to an observer— as a camera shot given to us as in a freeze frame

Charlotte Harris

or the blink of an eyelid brought to us so as to linger in the memory. If there were two train tracks, the train would take both of them

A Train with a Track in the Box

surely until the parallel tracks meet up ahead.

Although the train appears at a permanent standstill, it is still

Charlotte Harris

moving in the direction of travel, but a temporal standstill affects the temporal lobe in the brain and we believe it to be moving, even

A Train with a Track in the Box

at the speed of light. The other postulate is that it is just too heavy.'

'Well, I wouldn't look at it too long, especially if it is

moving. What about the tracks?'

'If you are riding a speed of light train, one to take you there and back again.

A Train with a Track in the Box

About the track, consult Einstein because you and I just haven't got time.

'And the train is made for

high-speed travel,' said Dr Huffman. 'So that clears up the mystery, OK?' and he smiled again, quizzically.

www.ingramcontent.com/pod-product-compliance
Lightning Source LLC
Chambersburg PA
CBHW021019180526
45163CB00005B/2027